Table of Contents

Preface 1
1. Why SOPs: What's the Big Deal? 3
2. The Importance of Standard Operating Procedures 12
3. The Process of Establishing Standard Operating Procedures 20
4. Questions to Consider 24
5. Determining the Tasks to Be Standardized 25
6. Establishing the Standard Operating Procedures to Be Used ... 31
7. Questions to Consider 38
8. Developing and Implementing the Procedures 45
9. Testing and Evaluating the Procedures 53
10. The Components of Standard Operating Procedures 59
References 69

Preface

This book was developed with the goal of assisting supervisors, managers, and water utility personnel in their desire to provide consistently professional responses to their customers.

A number of these critical employees inherited systems that were at best functioning at less than expected. Several of these organizations had workers who were never appropriately trained to perform their assigned duties, nor were they made aware of the connection between their work, the goals of the organization, and the satisfaction of those they serve.

This book provides information to assist agencies in properly developing and training their staff using procedures that focus on enhancing their knowledge and skill to improve their efficiency. Included are procedures to establish the optimal approach for standard operating procedure development, implementation, and sustainability.

The material in this book was developed based on industry standard guidelines and centered on regulatory requirements associated with the goal of effectively managing water distribution/wastewater collection systems, as well as appropriately addressing customer concerns.

Developing Effective Standard Operating Procedures
For Water Utilities

Kenneth C. Morgan, P.E.

outskirts
press

Developing Effective Standard Operating Procedures For Water Utilities
All Rights Reserved.
Copyright © 2021 Kenneth C. Morgan, PE.
v2.0

The opinions expressed in this manuscript are solely the opinions of the author and do not represent the opinions or thoughts of the publisher. The author has represented and warranted full ownership and/or legal right to publish all the materials in this book.

This book may not be reproduced, transmitted, or stored in whole or in part by any means, including graphic, electronic, or mechanical without the express written consent of the publisher except in the case of brief quotations embodied in critical articles and reviews.

Outskirts Press, Inc.
http://www.outskirtspress.com

ISBN: 978-1-9772-4376-8

Cover Photo © 2021 Kenneth C. Morgan, PE. All rights reserved - used with permission.

Outskirts Press and the "OP" logo are trademarks belonging to Outskirts Press, Inc.

PRINTED IN THE UNITED STATES OF AMERICA

CHAPTER 1

Why SOPs: What's the Big Deal?

The purpose of this book is to assist readers in understanding how to develop standard operating procedures (SOPs). A number of municipal agencies and/or utility service providers struggle to consistently give their customers effective service. The difficulties of managing an aging infrastructure using an aging and inconsistently trained workforce make the challenges even greater.

Agencies are also pursuing relevant ways in which their employees can be more capably efficient in the performance of their assigned duties. These desires can all be achieved with the development, implementation, and utilization of standard operating procedures. With the use of this material, wastewater collection/water distribution managers and supervisors have the ability to utilize techniques to capture processes that will assist them in standardizing work assignments.

When properly established and maintained, these procedures can provide

- a mechanism to train employees,
- a tool to evaluate the effectiveness of workers,

- a means of establishing efficiencies in the operation of the systems,
- a process to enhance the efforts of work teams,
- an approach to reliably align with regulatory requirements, and
- a proven system to accomplish the operational goals of the agency.

Additional topics covered in this book are intended to present some of the challenges associated with identifying the SOPs to be developed and selecting employees to assist in their development process, as well as the importance of establishing an effective schedule for their delivery.

Standard operating procedures are an essential tool to aid any utility in its desire to provide and/or enhance professional responses to the needs of its customers. When appropriately utilized, SOPs can positively impact the confidence that customers have in the agency's ability to maintain its systems.

Many municipal agencies and water service providers struggle with the reality of consistently developing their workforce in a manner that sustains the success of an effective operation. As seasoned workers retire, expertise and experience leave with them, and the value of their effort/skill to the organization is often lost. This is especially the case if the departing procedures have not been captured and/or standardized.

When the agency is faced with operational difficulties of any scale or size, having standard operating procedures allows the workforce to function appropriately with minimal disruption because of a reliable pattern for how the endeavor should be addressed. These procedures become a vital tool during times when crews or work teams have to be split up to accommodate changes in work assignments as a result of long-term schedule modifications.

WHY SOPs: WHAT'S THE BIG DEAL?

While training staff, responding to emergencies, or dealing with a modified workforce, the existence of standard operating procedures (SOPs) becomes critical. Making sure field crews and work teams understand the importance of their contribution to the success of the organization is important.

Being able to consistently introduce new employees to established approaches for completing assigned tasks offers safety-oriented mechanisms to evaluate their progress and guarantee critical aspects of their development. It is ultimately about efficiently developing employees to provide the best service to the customer while enhancing their confidence in their demonstrated skills.

A great number of utilities struggle to consistently provide their customers with reliably effective service. These struggles are made more evident during emergencies within their systems and/or difficulties with the responsiveness of their assets.

An agency's desire to provide a consistent, positive response to their customers can be achieved with the development, implementation, and utilization of standard operating procedures (SOPs).

The development and implementation of SOPs display the importance of the processes along with the value of the workers responsible for using them. When workers believe they are valued by their employer, their attitude about their assignments typically reflects this value.

An organized plan detailing the steps necessary to perform duties in alignment with the requirements of regulatory agencies reflects a compliance mindset as it relates to the work that needs to be done. This plan, when appropriately presented to all workers responsible for their execution, builds employee confidence.

There is a direct connection between consistent demonstration of the completion of assigned tasks on the part of the workforce established by the use of standard operating procedures and sustained satisfaction of the customer.

Customers appreciate being exposed to a knowledgeable worker with demonstrated skill when it comes to their service.

Apart from accessing the organization's website, one of the other interactions the customer has with the agency is when they engage an employee in response to an issue. Some of these interactions are related to engaging online and/or by phone, addressing water quality concerns and sewer lateral blockages, repairing water mains and service leaks, and addressing sanitary sewer overflows (SSOs), as well as informing residents of utility initiatives and programs.

A number of utilities do not regularly monitor their employee interactions with their customers. As a result, some employee responses are inconsistent and a number of them are often less than desirable. This lack of monitoring reflects a lack of training and accountability. It can also reflect the lack of a standard to be used to appropriately train workers, along with setting expectations with instruction.

When the productivity of workers is not regularly monitored or evaluated, the agency's ability to identify how well it is maintaining the critical assets of the systems cannot be determined in order to establish necessary improvements. This often leads to employees being allowed to work at their own pace or not be as productive as they are expected to be.

Employees who have not been held accountable for following specific instructions or those who have been used to doing things their way without any regard for the consequences will often struggle with the use of standard operating procedures.

The utility should set the expectation for employee development.

Expectations of this nature are based on organizational goals, regulatory requirements, and customer considerations. Along with this, the employee should have a desire for growth and opportunity. Their responses to the potential for growth can be clearly displayed by how they engage the customer. When employees are concerned enough to make sure each of their customer engagements is positive, the experience will demonstrate value on both sides.

The Use of SOPs for "Best in Class" Utilities

There is much talk across the industry about agencies being "Best in Class" or using "Best Operational Practices". The use of standard operating procedures to accomplish an agency's desire to be "Best in Class" helps by reducing

- additional costs associated with inconsistent and/or inadequate responses,
- the additional loss of the product/service provided because of inefficiencies,
- the time it takes to restore the service to the customer in a manner they expect,
- the amount of time wasted using outdated and/or inconsistent processes that rarely work, and
- the confusion employees experience when there are no standards.

Inconsistent and/or inadequate responses almost always cost the utility more money to maintain the same level of service requested. On a time and material basis alone, inconsistent activity on the part of the crews performing maintenance, operation, repair, rehab, or replacements can be more expensive and will make it more difficult to properly determine the budget necessary to keep assets functioning.

DEVELOPING EFFECTIVE STANDARD
OPERATING PROCEDURES FOR WATER UTILITIES

Assets that are inconsistently maintained tend to fail at a faster rate than those provided with a regular schedule of attention. When isolating an area to respond to a water quality issue, an SSO, or a water main break, inconsistent responses in most cases will result in a greater loss of water or service along with an increased area of customers negatively impacted.

Obviously, when a greater number of customers are negatively impacted than hoped for because of inconsistent responses, the time it takes to restore service after an issue and/or response will increase as well. More time, more expenses.

Another contributor to the work completed being more expensive and taking more time is when there is confusion on the part of the workers because of the lack of any standards for fulfilling the assigned tasks.

The effective use of SOPs encourages the worker to acquire and maintain a level of professionalism demonstrated by the way they complete assignments and/or projects. When employees believe they are a critical part of the success of the organization, their responses to their assigned tasks reflect it.

SOPs for Emergency Response Planning

When up-to-date standard operating procedures are in place during emergency response planning and/or responses, they offer dependable processes to reduce the potential for employee error.

They provide staffing flexibility to allow for the workforce to be divided as needed without concerns of placing some workers in a vulnerable position based on their diminished skill level compared to others.

WHY SOPS: WHAT'S THE BIG DEAL?

Responding to emergencies presents a greater challenge to how work is assigned and completed when unequally skilled workers are negatively impacted by the incidents. Employees who have not been appropriately trained will place the organization in a vulnerable position during these times. An emergency will often expose what skill level really exists among responding workers. It will either reveal the value of the training and preparedness or expose the need for it.

If commonly known and used standard operating procedures are not in place during these critical times, the potential for using outdated processes can occur, resulting in inconsistent approaches that cause or further the difficulties associated with the problems. When this happens, the more knowledgeable workers will often be spread thin and overworked, and an increase in the potential for injuries will follow. Because of the inconsistent knowledge of the workers, certain important assets or associated tasks may be delayed, rescheduled, or even allowed to run to failure.

SOPs for Continuity of Operation Planning/Processing

Current SOPs provide a mechanism for workers to keep assets reliably functional during the "continuity of operation" effort. Negative workforce impacts can be mitigated by staff who do not normally perform selected tasks because of the availability of the SOPs. This can be accomplished by using the SOPs as a cross-training tool.

In contrast, outdated or absent SOPs during continuity of operation periods will more than likely result in increased downtimes, potentially avoidable equipment failures, and an extended loss of service. In these scenarios, there is a greater possibility of introducing a contaminant into the system, as well as any related difficulties. The lack of current SOPs can often expand the area negatively impacted by a sanitary sewer overflow at a critical manhole or lift station.

DEVELOPING EFFECTIVE STANDARD OPERATING PROCEDURES FOR WATER UTILITIES

Restoring effective service to the system, or a critical portion of it, after a major event can be stifled by the lack of SOPs. Proven and practiced SOPs help to get the system up and running at a faster rate than if they were not available or not used. Not having SOPs in place causes difficulties in the restoration of service process resulting in workers using inconsistent, fragmented instructions or being required to provide limited responses.

Limited responses during emergencies or when service is minimally maintained through a major event cannot be performed to any degree of effectiveness without SOPs. These vital processes, when available, offer the most appropriate way to bring the assets and/or systems back on line as well as up to full operation.

Questions to Consider

Can you identify any challenges your agency has faced because of the lack of standard operating procedures or the existence of outdated ones?

Identify some of the benefits of having current SOPs used in the maintenance, operation, repair, rehab, and replacement of your assets.

What messages are being presented to employees when there are no structured procedures in place to explain how the work should be done?

Identify some of the benefits of developing and enhancing employee confidence by the use of SOPs.

CHAPTER 2

The Importance of Standard Operating Procedures

STANDARD OPERATING PROCEDURES often set the tone for the effectiveness of any agency. Across all industries, the entities that have standardized their procedures are usually the leaders in cost-effectiveness, efficiency, productivity, and safety.
A number of sewer/water service providers have rarely used a consistent approach to developing their staff and have had to deal with inconsistencies in how their employees engage their customers, how the work is completed, and how they measure their success, as well as a myriad of personnel issues, as a result of these productivity challenges.

Managers and supervisors who have inherited low-performing or nonproductive workers can use SOPs to establish the expectancies they desire to aid in bringing their organization within the range of those high-achieving utilities. It starts with the desire to be better. Managers and supervisors interested in being a part of a better-performing agency will dedicate time and energy to developing their workers.

A manager/supervisor who wants more out of his or her workers will often give more to them in the form of

THE IMPORTANCE OF STANDARD OPERATING PROCEDURES

- consistent positive engagements,
- an organized training plan,
- agreed upon expectations, and
- continuous monitoring to guarantee success.

Consistent positive engagements between supervisors and their workers display a genuine interest in the development of each employee. During these engagements, employee knowledge can be evaluated, assignments can be discussed, and safety concerns can be addressed.

The use of SOPs during some of these discussions can help to establish an organized training plan for the worker. This plan should include agreed upon expectations, a timeline to assess each worker's progress, and necessary requirements for improvement if the expectations are not met.

In order to be of any real value to the employee and the organization, the worker's progress should be regularly monitored and discussed.

SOPs are an essential part of the platform necessary to prepare workers to perform their assigned tasks, along with potentially becoming the future leaders of the organization.

A critical part of the success of using SOPs is to determine what processes are selected to be standardized. If every process is standardized, employees may become bogged down by being forced into a rigid methodology that does not allow flexibility where necessary.

SOPs provide a mechanism to consistently train and evaluate employees.

Responsible municipal agencies and water/sewer service providers are often pursuing ways in which their employees can be more capably efficient in the performance of their assigned duties. SOPs enhance the responsiveness of field crews and work teams by making their tasks more uniform. This uniformity increases their understanding of how the assets they are responsible for function.

How does an agency identify what should be standardized?

Any assignment or task performed by a number of workers several times during the course of the day, week, or month should have a standard procedure associated with it.

Later in this book, the specifics of determining what tasks should be standardized will be presented in more detail. In light of that, care should be taken to select tasks that reflect the professionalism desired for the assignments given.

Standard operating procedures

- assist in identifying crew sizes and equipment needs,
- can be used to consistently develop employees, and
- determine the reliability of field crews.

Identification of crew sizes and equipment needs considers the established service levels of the department/division. Another important consideration should be the anticipated size of the system and/or service area.

The anticipated size of the system incorporates the existing system size, the potential for the projected increase/decrease, and/or the expected rate of expansion or reduction. Some utilities have or may be in the process of absorbing another system into their service area. This addition can create difficulties in determining the optimal response times and processes to address customer concerns.

THE IMPORTANCE OF STANDARD OPERATING PROCEDURES

In contrast, a reduction in the service area or in the number of customers receiving service can create water quality problems that will generate added activity to the existing duties of the field workers. The larger negative impact of this is the result of a high-volume customer moving out of a community, greatly reducing the amount of water and/or sewer service supplied to an area.

This directly impacts the time it may take workers to respond to the concerns of the customer or to accomplish needed tasks in a timely manner. It will reflect the level of expertise and training necessary for an effective workforce.

A consistent employee development process involves evaluating what the workers currently know, standardizing what they are taught, and modifying training as necessary.

What are the essential steps for evaluating the existing levels of knowledge of field crews and work teams?

First, it is necessary to identify the criteria for assessing the knowledge of field crews. This effort should be started by determining what the workers are responsible for by job function. Primary job functions, secondary job duties, and shared assignments should be identified as part of this process.

Secondly, it is vital to determine the worker's level of understanding about the duties they are responsible for. This can be done a number of ways. It can be accomplished in a classroom setting by allowing workers to demonstrate their knowledge in an assessment questionnaire or tool. The questions should be based on the workers' ability to show their knowledge of what is expected as a result of their job duties.

Another way to determine the existing knowledge level of field crews and work teams is to evaluate their performance of selected duties in the field.

The approach the workers take when certain steps are performed, the efficiency displayed, and the quality of the effort exerted should all be evaluated to determine their level of demonstrated knowledge or what is understood.

A combination of the two previous approaches can offer a balanced assessment process as well. This takes into consideration the reality that some workers are visual learners and doers as opposed to being able to demonstrate their knowledge effectively in a classroom setting with a written assessment tool.

This also presents critical information that will assist the agency in identifying development and training content to fill learning gaps and thus enhance the knowledge/skill/capabilities of field crews and work teams.

The demonstrated knowledge of field crews and work teams has a direct impact on the reliability of these employees. The more they know, the more they are able to confidently demonstrate it. The reliability of field crews and work teams is often dependent upon their knowledge of industry standard responses, their awareness of the functions and location of system assets, as well as their exposure to and understanding of current and/or new technologies designed to enhance responses in the water industry.

Field crews and work teams aware of current industry processes will often demonstrate this by their ability to be efficient in the completion of their assignments, along with being educated in the connection between regulations and their duties. This displayed awareness directly benefits their customer interaction and raises the level of appreciation.

THE IMPORTANCE OF STANDARD OPERATING PROCEDURES

Knowing what to do is fine, but if the worker does not know how the asset fits into the processes of providing a quality product to the customer, their desire to be effective will be hindered. Understanding the functions of assets—and how they interact with adjacent components and other related resources—helps by being able to keep things working during normal operations, but more importantly, it offers valuable tools when responding to emergencies.

Being aware of the location of assets cannot be overstated as well. The asset's location can determine the critical approach required to resolve an issue. If the location of the asset is unknown, it can be the difference between negatively impacting a small subdivision versus a larger part of the service area. Asset location can also be beneficial to identifying the most appropriate way to get customers back in service after a major repair and/or water quality event.

Exposing field crews and work teams to current or new industry technologies advances their awareness of more innovative ways to perform their assignments. It often broadens their understanding of the importance of their role in providing a quality product to their customers. New technologies can often pique a worker's interest in determining a better way to do their job.

An understanding of the latest technology available includes being aware of appropriate rehabilitation, repair, and/or replacement techniques; selecting the right application for the solution desired; and determining if staff is prepared to use the technology evaluated.

In addition to the exposure of new technology, field crews' and work teams' reliability can be increased by consistent operator engagement and training.

Evaluating the level of expertise and training of the staff considers the effectiveness of the current training, the consistency of the training with current operational practices, and an ongoing assessment and modification process.

When SOPs are used, crew sizes and equipment needs can be determined more accurately, evaluated more consistently, and adjusted and/or modified more appropriately.

Standard operating procedures can be used to steadily develop employees by establishing a standard approach to training, teaching a consistent process and/or procedure, and assisting in the development of modifications in the proper manner.

These standardized instructions help determine the reliability of field crews and work teams by providing a model for the work to be done, being a continuous improvement tool, and offering a process modification mechanism.

The ultimate benefactors are the customers who rely on their water/sewer service provider to effectively train their personnel, to be aware of the requirements of regulatory agencies that impact their lives, to cost-effectively manage the assets they are responsible for, and to appropriately communicate critical information when necessary.

A well-run water/sewer agency will often demonstrate their effectiveness by being responsive to their customers, in collaboration with the regulators that manage them, and maintaining their assets, utilizing the latest reliable technology the industry offers to align with organizational goals.

THE IMPORTANCE OF STANDARD OPERATING PROCEDURES

Questions to Consider

In what ways do you believe SOPs will benefit your agency?

How do your supervisors/managers currently set expectations for your field crews and work teams?

Does your agency have effective standard operating procedures in place, and are they being consistently used? If not, why?

Does your agency use SOPs as an assessment, development, and training tool for your field crews and work teams? If so, how?

How does your agency measure the effectiveness of your field crews and work teams?

CHAPTER 3

The Process of Establishing Standard Operating Procedures

IN THE PROCESS of producing SOPs, some organizations have taken the procedures of another agency, modified them sparingly, introduced them to their workers as the next "great idea" without allowing them the benefit of participating in the development effort, and were strangely surprised when what was presented was not well received nor consistently used.

"It is a matter of ownership."

The difficulty some agencies face when they bring in someone else's SOPs—with the hope of modifying them without the engagement of the staff responsible for using them—says a lot about their opinion of their workers. Whether stated or not, this implies a lack of confidence in the workers' ability to participate in this important process. Those responsible for the use of the SOPs have more at stake with their being correct, executable, and consistent than anyone else. For this reason, when it comes to the successful development, implementation, and continued use of SOPs, the workers should be actively involved in the process.

The involvement of these critical workers should be based on their

THE PROCESS OF ESTABLISHING STANDARD OPERATING PROCEDURES

availability, capability, and interest; however, by all means they should be engaged. Some organizations have had draft copies of SOPs presented to the field crews and work teams for their review and initial acceptance. This process offers the impacted workers an opportunity to see how their assigned tasks may be changed prior to the documents being finalized. This approach suggests to the workers that their opinions have value and are requested.

There are parts of the SOP development process that are universal to all organizations; however, the more critical parts are unique to the agency along with their assets, and if not treated as such, the agency can often lose an opportunity to enhance the effectiveness of their workforce.

One of the primary challenges associated with the effective utilization of SOPs is to convince the employees responsible for their execution to accept their value to the organization and ultimately to them. The desired success cannot be achieved in earnest if these vital workers are not appropriately engaged in the development and implementation processes.

Developing SOPs without the full, continuous engagement of those responsible for using them can be done, but it will not demonstrate the value the organization places on the worker nor an expressed commitment to their success. One of the keys to this engagement is centered on identifying field workers with the potential to lead and/ or those who are leaders.

The more these leaders are engaged in the development, implementation, and modification of the SOPs, the greater their acceptance will be. The greater their acceptance, the better the responses of the impacted workers. As the responses of the workers improve, the efficiency of the operation will improve. When the operational efficiency is improved, the productivity increases, which will ultimately be observed by the customers and potentially result in a higher approval rating.

Establishing the SOP development process is critical to the success of the execution of assigned tasks.

Valued employees realize the development of SOPs is vital to their personal success. When this development is standardized by using SOPs, it provides a consistent mechanism to guarantee the proper execution of assigned tasks.

System realities play a crucial role in identifying the appropriate SOPs. Component and system complexities are part of these realities and must be factored into what training is eventually determined, as well as how they are to be used to the benefit of the organization. During the course of SOP development, knowing that a particular step should precede another one for a specific asset can be the difference between isolating the right area and negatively impacting a greater number of customers.

Knowing the interaction between the system complexities and the skill level of the impacted staff (current and proposed), along with the desire to perform these tasks in compliance with the applicable regulatory requirements, is vital to any utility's success.

The SOP development process should consider the accuracy of the system's complexity, the desired skill level of the impacted workers, and the concern for safety, as well as compliance with all applicable requirements.

The system's complexity can impact SOPs by identifying the number of processes to be developed, contributing to the determination of the level of staff involvement, and encouraging a safety-focused approach to their development. This shows how important it is to

THE PROCESS OF ESTABLISHING STANDARD OPERATING PROCEDURES

coordinate the SOPs with how the system should be run, as well as balancing the understanding of those responsible for it.

The desired skill level of the workers can impact the development of SOPs by having to consider whether to oversimplify the procedures or not. The SOP development process can also assist in providing a mechanism for breaking field crew/worker bad habits or improving inefficient procedures. Addressing these issues up front with the impacted workers goes a long way to achieving the desired expectation for the effort.

The concern for safety impacts the SOP development process by showing safety as an integral part of the document instead of a secondary component. In this approach, safety should be displayed prominently in the execution of the assigned tasks. References to established safety equipment, materials, and tools should be added to the SOP to confirm their importance.

The desire of the agency to comply with all applicable requirements should impact the SOPs by potentially identifying the specific requirements as part of the document. Another potential approach is to display the value of compliance within the body of the SOP. The connection between the SOP, compliance, and effective service should be a consideration within the document. Care must be taken when considering what to include in the document to make sure it is not too wordy or extremely long. When one of the goals of the agency is to show the connection between the SOPs and important regulations and/or requirements, it further enforces their value and potentially the accountability expected.

The process of establishing SOPs determines the tasks to be standardized, builds the procedures to be used, develops and produces the plan of action, implements the use of the operation, and allows for necessary modifications.

Questions to Consider

Identify some of the issues associated with a system's complexity that will impact the SOPs developed.

Describe the impact improperly trained workers will have on the development of SOPs.

Describe potential ways to list compliance information within the body of the SOP without it being too wordy.

Describe how safety should be incorporated into the SOPs.

CHAPTER 4

Determining the Tasks to Be Standardized

SOME AGENCIES HAVE a procedure for almost everything. This approach can have a negative effect on the value of SOPs, especially when the previous procedures were not standardized. The difficulty is that these unstandardized processes are performed various ways by different field crews and/or work teams without consistency. The goal should be to determine specific tasks to be standardized based on established criteria.

Identification of criteria essential for establishing which procedures get standardized is an important part of the process. The criteria should be determined by the goals of the organization along with the capacity of the impacted workers to reasonably complete the tasks.

The essence of the questions asked and eventually the criteria should focus on making the responses to assigned tasks more efficient, protecting the sustained operation and maintenance of assets, as well as presenting the best demonstration of knowledgeable employees to customers.

It is vitally important to have the criteria established before identifying the list of potential standards. This reduces the time spent going back and forth deciding whether a procedure is selected or not because of ever-changing determining factors.

Components of the identified criteria should include the importance of the SOP to the overall system or a critical asset. Part of the criteria should reflect a response to compliance with a regulation (where applicable), or an established system goal. Some criteria may not be directly related to these items but may be necessary to support their accomplishment. The criteria selected should be specific to the agency, consistent with the plan to gain efficiencies, and appropriate enough to properly standardize the task(s).

There are few greater frustrations for workers than to be required to perform assignments that they are not capable of performing or that they cannot complete with the desired level of success.

This reduces the confidence of the employees and their work teams. It also displays a lack of commitment to the workers' success on the part of those requiring the "unrealistic goals." This can increase the potential for disgruntled and underperforming workers.

With nonstandard procedures, new employees may experience a loss of confidence based on having to relearn something they thought they knew when transferred to another crew instead of being able to further practice a more standardized approach.

The capacity of the impacted workers to reasonably complete the standardized tasks should be a major part of the development process. The negative impact of seasonal difficulties and/or system anomalies can be reduced by the existence of standardized procedures to be used when needed. Addressing these departures from normal procedures within the body of the document can often lengthen it or complicate its understanding. For this reason, whenever added, the additional processes should be notated as such within the standard.

DETERMINING THE TASKS TO BE STANDARDIZED

Standardized tasks should be directly related to the agency's goals. This connection should be clearly identified in a way that allows the worker to understand that when assignments are completed properly, their importance and the value of their organization increase.

Addressing employee concerns regarding the process up front can go a long way to reducing the potential for pushback or opposition to the desired success. This provides an opportunity to identify possible "pinch points" that can shipwreck the process or extend the anticipated schedule of completion. The importance of these discussions cannot be overstated.

Another advantage of having these up-front discussions is that they create a platform for the employees' interest to be evaluated, identifying probable SOP development participants. It may even give an early peek at those who are posturing themselves against the effort.

Determining the tasks to be standardized involves selecting tasks where efficiencies can be gained, understanding the skills of those who will be performing the assignments, and being aware of the latest technology to perform the work.

Potential areas where efficiencies can be gained are as follows:

- Preparing for a repair
- Selecting the right equipment/material
- Setting up the travel zone
- Staging the work zone
- Initially engaging customers
- Isolating an impacted area
- Completing a repair
- Restoring service to customers
- Backfilling and compaction
- Surface restoration

Selecting tasks where efficiencies can be gained involves establishing criteria for SOP development as mentioned above, as well as providing a clear understanding of the selected tasks. Another aspect involves determining areas for efficiencies within the essence of the procedure. This relates to assessing the steps within the procedure to identify areas where the time to perform or complete certain portions of the document may be positively adjusted.

It is more than a matter of just selecting certain tasks. The process should also involve an understanding of these tasks, why they were selected, and the intent to develop and/or enhance the productivity of the worker/team.

The approach used for training, along with the material employed, can often be dictated by the workers' skill level. The process and length of the training effort should be established to gain the optimum benefit for the impacted workers. For the greatest value, achieving the training objectives should also be incorporated into the employees' goals.

Once the list has been identified, the prioritization of the selected tasks should be accomplished to rank them in order of importance based on the same criteria. This is one of the best approaches for establishing useful SOPs. Some SOPs are prioritized based on their value to the organization, a particular area, an asset, and/or associated components.

The prioritization process is usually specific to the agency and has more to do with the complexity of the system, the way the organization functions, its culture, and the value it places on customer interactions.

These areas interdependently and collectively shape what an organization deems as more important than something else. An example of

DETERMINING THE TASKS TO BE STANDARDIZED

this relates to the interaction between the organization's culture and the regulatory requirements.

Knowledge and awareness of regulatory requirements should be clearly understood by an organization and its employees all the way down to the newly hired laborer; however, if these requirements have not been prioritized within the culture of the organization, they will not be adhered to properly. For this reason, a number of water/sewer service agencies find themselves in violation without realistically acknowledging how it happened.

Prioritizing the SOPs to be developed assists the agency in establishing a reasonable schedule to produce procedures based on their importance to the workers, the system, and the organization.

Questions to Consider

Identify a list of tasks that should be standardized and why.

List some of the efficiencies gained by standardizing procedures.

Identify a list of system goals to be addressed by SOPs.

Identify criteria essential for determining what tasks should have an SOP associated with it.

What are some of the ways to prioritize the development of SOPs?

CHAPTER 5

Establishing the Standard Operating Procedures to Be Used

IN ORDER FOR water/sewer service providers to consistently experience the success of SOPs, their instructions must be executable and measurable. If the tasks explained by the SOP cannot be performed in a reasonable amount of time without difficulty, the procedures lose their value.

Workers' ability to reasonably execute the SOPs has more to do with being able to understand the tasks requested of them to complete. This understanding is often based on the exposure these workers have had to the appropriate way of completing assignments.

When a standardized process of completing specific tasks is presented to the workers prior to their being asked to perform them, their ability to know and understand what is required of them is often enhanced.

Regular reviews of work processes are an important part of guaranteeing the consistency of the work being done. These reviews should take place as a means of evaluating the quality of the work, be used as a training tool to enhance productivity, and be a mechanism to exact discipline when the requirements have been violated.

DEVELOPING EFFECTIVE STANDARD
OPERATING PROCEDURES FOR WATER UTILITIES

These reviews have a greater importance when they are done using the same set of standardized processes provided to the workers prior to their receiving the assignments.

It is an unfair and frustrating challenge to hold a field crew and/or work team responsible for completing work when they have not been exposed to the proper way of performing it. This exposure should obviously include being allowed a reasonable amount of time to finish the assignment without having to rush, skip critical steps, or risk the safety of the crew and the environment in which the work is being done in.

If the workers using the SOPs struggle to understand what must be done, it is highly unlikely the tasks will be completed properly. If they are completed, it will not be without difficulty.

These difficulties could result in accidents and/or injuries as well— once again, potentially reducing the confidence of the worker, the team, and ultimately that which the customer may have in them.

In order to confirm any success, it must be measurable. If it is not measurable, the agency will not be able to show, with confidence, the effective completion of the assigned tasks. This measurement can be accomplished consistently when the procedures being evaluated are standardized.

The process of measuring the time it should take to complete an assignment can be based on establishing an initial period for completion, evaluating it against the actual time it took to finish the task, and adjusting where necessary.

In order to more appropriately balance the reasonable time, consideration should be given to the various situations potentially faced during the completion of the assignment. An example of this considers the time it might take to remove asphalt versus concrete when repairing a water or sewer main break. In most situations, the removal of concrete will require a bit more time to accomplish. This example can be extended to also consider the time of repair in a residential street versus a busy thoroughfare.

Reviewing the time allotted to perform water/sewer main breaks in different environments should lead to a time being selected that considers an average between the two scenarios. The same process should be considered in similar situations where there are different types of materials used and/or variations in the environments to complete the work. Another major factor is the part of the country or region. Based on this, there may be differences in the time it may take to complete the assignment during the spring/summer versus the winter.

A reasonable approach is to initially establish a time and review the work done over a period of up to three to six months or so, making sure a representative number of these assignments have been completed within the established time frame. This affords the ability to review the work, identify similarities, adjust times where necessary, and reestablish a reasonable time based on the recent assessment.

When this reasonable time has been identified, it becomes vital to allow that time to be the standard until the SOP is set for review, modification, and/or revision.

Once finalized, data can be captured to note the average times for task completion, but those times should only be changed with the same frequency that the SOP is changed, unless there are glaring differences. The frequency of when and how often the SOPs should be changed is covered later in this document.

Establishing SOPs consists of identifying procedures that will be user-friendly, determining processes that are accurate and concise, selecting approaches where the possibility of modifications to the procedure is reduced, and placing an emphasis on safety within the body of the document.

User-friendly SOPs reduce the anxiety of those responsible for their execution. SOPs are user-friendly when they are easy to teach and train, easy to understand, easy to follow, and easy to modify when necessary. Reviewing the procedures should not confuse or stress out those responsible for using them to perform their work. They should allow a rational worker to review them and then go about completing the assignment without having to come back to verify what needs to be done next. Without this clarity, the use of an ambiguous SOP will definitely hinder any amount of efficiency desired in the completion of the work. For this reason, taking an ample amount of time for training makes a difference.

To some, the material may be new. To others, it may be a dull refresher. However, necessary attention should be given to guarantee the workers' understanding of the tasks at hand, as well as how to complete them safely.

The accuracy of the procedures with industry-standard approaches is vital to their being respected by those using them. It offers a better platform for their consistent execution and proper measurement.

Accurate and concise SOPs should be specific enough to be understood without error or misinterpretation, executable without difficulty, developed for operational efficiency, and consistent with the theme of the procedure.

As mentioned before, the more the SOP is understood, the easier it will be to follow. One agency used what was called "the new worker test." They gave the SOP to a new employee and asked him or her to complete the procedures based on their understanding of what was being requested of them. If the employee was able to complete the tasks with little to no difficulty, it was considered a success.

This may not work for all standards, particularly the more complex ones; however, it is a reasonable approach to be considered for the straightforward ones, particularly with smaller organizations.

If SOPs are not developed with a consideration for operational efficiency in mind, they will not be achieved. This is not to say the SOP should be short-circuited or some of the steps reduced. It should be developed with the intent of identifying ways to complete the tasks safely in the most efficient manner possible. Safety should not be sacrificed to gain efficiency and vice versa. There should be a balance between the two, resulting in a reasonable process to finish the assignment.

The steps identified in the body of the SOP should be consistent with the theme or goal of the procedure. If there is a disconnect between the goal of the SOP and the steps provided in the document, confusion will ensue. The importance of this matter will be covered in more detail later in this document.

There should also be a level of awareness of current technologies when developing SOPs to show reliability within the agency, along with an effort to keep employees informed. Not all new technologies can be integrated into the proper operation, maintenance, and management of assets. However, being aware of these technologies offers the best ability to know which ones may bring value versus the ones that may just be "pipe dreams" (pun intended).

The SOPs developed should be established with an emphasis on the safety of those performing the assignments. These safety measures should incorporate the staff's understanding of the task(s) and the appropriate operation of the equipment used. Another important factor is the environment in which the work will be accomplished. Knowing the hazards associated with the environment where the work is done factors into how success will be achieved.

Consideration for the safety of the personnel performing the tasks involves their knowing what is at stake from the standpoint of safety. This is reflected in the specifics of the steps to be done with an emphasis on completing them in a safe manner. It should convey an awareness of the environment in which the work is to be done specific to understanding the hazards from all aspects of the job.

When equipment is used in the execution of the job, knowing the potential for accident/injury should be paramount.

Precautions should be taken and even discussed before the work begins so that all involved are well aware of the potential safety challenges associated with the use of small hand tools, select materials, adjacent utilities, and heavy equipment.

Consideration should be given to potentially adding these concerns to the SOP and/or referencing them in an associated document.

As important as the safety of the staff is, so is the importance of the customers' safety. Certain jobs are set in environments where the public could be placed in harm's way based on the nature of the work or some related issue. An example of this is a gas line and sanitary sewer main adjacent to the repair work on a water distribution main.

The impact of the proximity and hazards of other utilities should be accounted for wherever they exist. Knowing the location of these other utilities prior to excavating is just one part of accounting for them. A process for understanding their impact on the completion of the work can be of value as well.

ESTABLISHING THE STANDARD OPERATING PROCEDURES TO BE USED

Vehicular traffic can endanger the work team, as well as the driver, if the area of the work is not secure. In rare instances, even when the site is secure, inattentive drivers or pedestrians may venture recklessly into the work zone and cause harm to an employee, another vital asset, or themselves.

Pedestrian traffic may place people in harm's way as a result of some jobs being performed in heavy traffic areas or locations where there is a higher volume of foot traffic, such as near a school or in a strip mall, downtown area, or business district. Work performed in these areas must be planned with these complexities in mind. Information reflecting the challenges of working in these areas should be discussed but not necessarily placed in the body of an SOP.

Work should always be completed with consideration for the safety of the customers directly impacted. Whether the job is completely on their property or in a part of their facility, or adjacent to their yard, etc., a demonstration of the safe performance of assigned tasks should potentially be included in the SOP. This safety information should emphasize its importance to the physical labor and the quality of the water, as well as consider the potential for their emotional safety where applicable.

Questions to Consider

Identify considerations to be used to make SOPs executable and explain why for each.

Determine how to evaluate the time it takes to perform standardized tasks identified as part of an SOP.

Describe the benefits and challenges with establishing an accurate and concise SOP for installing a water service line.

List the challenges associated with the SOPs being developed by someone who is not responsible for using them.

CHAPTER 6

Developing and Producing Procedures

DEVELOPING AND PRODUCING useful SOPs should consider the involvement of knowledgeable workers who will be performing the tasks. Their involvement reflects the confidence their superiors have in their ability and expertise.

This confidence often will be played out in the level of professionalism these employees demonstrate when completing their regularly assigned duties, as well as when engaging with customers. For this reason, these should be workers with a desire to be an active part of the development of SOPs, not only to develop the documents but also to set the best examples for how to execute them.

The involvement of knowledgeable workers also sends a message to less experienced employees that rewards of involvement in other important duties are given to those with demonstrated abilities and leadership. These workers are respected for what they know as well as for the professionalism they display in accomplishing their assigned duties.

For the sake of fairness, a methodology should be established and used to determine the subject-matter employees who will participate in the SOP development effort. When selecting employees, managers should consider using workers reasonably knowledgeable in the area of the SOP execution, respected staff without disciplinary issues, good team members who are easy to work with, and employees with a desire to learn and/or teach.

Having criteria for employee involvement sets the pace not only for what is expected to develop the standards, but also for what examples are used to maintain them. It also shows underachieving workers that lackluster effort will not be rewarded with special assignments or additional activities.

The leadership of these workers can be instrumental in the acceptance of the SOPs. Their involvement in all phases of the development, implementation, and maintenance of these critical standards aids in the sustainability of the desired service to customers.

These employees should be known for completing their work on schedule and in a timely manner. This is the same approach necessary to produce and implement SOPs in a time-sensitive fashion.

In the unfortunate reality that there may not be enough leaders within the work group to cover all areas of SOP development, steps should be taken to develop other employees who want more from their employer as well as want to give more. This exercise will offer an opportunity for these workers to grow and increase their exposure to additional promotional opportunities.

Developing and producing SOPs include identifying a standard process for SOP development, determining a consistent time frame dedicated to their production, specifying a selected time for review and revision, and presenting the developed SOP in a timely manner.

DEVELOPING AND PRODUCING PROCEDURES

The standard process selected should be based on what works best for the development team. After the list of SOPs has been prioritized, one approach is to develop the SOPs one at a time. This involves developing the goal and/or purpose statement of the SOP, the critical steps to follow, and any other pertinent information desired in the document. Once each SOP has been completed, a schedule for the review process can be established.

Another approach to consider involves identifying a certain number of the top SOPs, say three to five, or whatever number the team feels confident in tackling; developing the goal/purpose statement of these identified SOPs; reviewing them to a point of initial acceptance; and then moving on to repeat the same process with the remaining components of the selected SOPs until they are all complete. Once complete, the team can move to the next established number of SOPs and repeat the process until they are all finished.

Yet another approach consists of dividing the development team into small subgroups. Each subgroup can produce a certain number of SOPs, using one of the processes mentioned above and then presenting their developed standards to the other subgroups for review and/or revision.

There may be other developmental procedures that may work better for the agency involved; however, no matter what approach is used, it is very important to keep the groups on task and on schedule.

There has to be a level of accountability that guarantees the development/production of these essential documents within the time frame selected.

In order to gain the greatest value to the organization, the development and production of the SOPs should be accomplished within a reasonable amount of time. In establishing a workable schedule for the development of the SOPs, the priorities and complexity of the system, as well as the knowledge level of the participants, can directly impact the process and should be considered. A reasonable schedule enhances the confidence of the employees involved and directly benefits the system they were designed for.

If the SOPs are not developed and implemented in a timely manner, they can lose their value or importance to the employees and the organization. Many great initiatives have been scrubbed because their value was diminished by a lax schedule or lackluster support from the managers and those responsible for their execution.

The emphasis should be on producing SOPs consistent with the priorities of the system and in alignment with the goals of the organization. To guarantee the success of the SOPs and eventually the organization, the produced tasks should be tested prior to final implementation.

Developing and producing the procedures entail involving some of the personnel that will be responsible for their execution. The establishment of a schedule for the development and completion of the draft documents, as well as their testing and final acceptance to guarantee the success of the effort, should be prioritized among the other assignments of these workers. If this effort is not prioritized, it will never get done. Workers are always busy getting things done; however, what is prioritized gets done!

The same approach reliable employees use to complete their day-to-day assigned tasks is necessary to produce and implement SOPs in a reasonable amount of time.

In another shameless attempt to emphasize the importance of involving staff, doing so presents a greater potential for employees to buy into the effort. It also can generate a higher level of their understanding of the SOP.

DEVELOPING AND PRODUCING PROCEDURES

Staff involvement in the completion, implementation, and continued use of an SOP should have a positive impact on the efficiency of the system and the organization. To those directly engaged, the process enhances the focus the workers have on the standard.

If the employees of the organization do not have the capacity to develop and implement their SOPs in a timely manner, consideration should be given to acquiring the services of an agency and/or consultant to assist in the effort.

The produced tasks should be evaluated and tested prior to final implementation to guarantee their success. Testing and evaluating the procedures involve making sure they are achievable, reviewing the thoroughness of the steps within the documents, and identifying the benefits of the processes in a manner that reduces the number of SOP modifications. A schedule for this part of the process is essential as well.

Questions to Consider

Identify the benefits and challenges of developing SOPs in a timely manner.

Identify a list of SOPs critical for maintaining water quality within the distribution system.

Describe steps essential to properly evaluate the thoroughness of an SOP.

Identify additional criteria to be considered when selecting which employees will participate in the SOP development process.

CHAPTER 7

Implementing the Procedures

AGENCIES INTENDING TO introduce new initiatives and/or programs can often face some of the challenges associated with a workforce that does not accept change that well. Some of these workers may be seasoned employees who have experienced frustrations over the years, were passed over for promotion, and/or have become cynical about the plans of the organization.

In too many cases, the power of the culture that was allowed to be established over the years by the disgruntled worker or an underperforming work group is greatly underestimated. Observing it from the sidelines often causes concern about who was really in charge within the work group, based on what things were allowed to exist or what tasks were or were not completed.

The impact of one unhappy employee can create havoc in a work group trying to find its identity or attempting to transition from one type of manager/supervisor to another. Typically, a new supervisor or manager will change procedures and/or requirements with the intent of altering the group culture. These desired alterations are often met with resistance. Accepting the behavior of an underperforming team can stifle any desired progress if not addressed when noticed.

DEVELOPING EFFECTIVE STANDARD
OPERATING PROCEDURES FOR WATER UTILITIES

The contention can be made even worse if the particular disgruntled worker has been in the group longer than the new supervisor/manager or if the nonproductive workers have not been consistently held accountable. This often results in turf wars that should not exist but do because of the lack of clear direction or, as mentioned before, if the group is trying to establish its identity.

Another struggle can potentially arise if the supervisor/manager was promoted from within the ranks and that person has had a not-so-admirable history within the work group or organization. In many cases, when trying to move the group forward, some workers will remind the "new supervisor/manager" of the way things used to be with the intent of hindering their desire to move beyond the past.

Employees who just want to do their jobs are usually caught in the middle, often being forced to "choose a side" or "declare loyalty" or suffer the wrath of the discontented employee(s). It is truly amazing how these individuals are allowed to control their work environments without officially being in a position of authority.

Engaging these disgruntled and/or underperforming workers on the front end of the process is essential, and if this is not done in a timely manner, it can derail the desired success. These workers should be engaged by allowing them to express their concerns, addressing their issues upfront, and agreeing to an acceptable resolution moving forward. If an amicable resolution cannot be established, the supervisor must exercise his or her authority to direct the actions requested.

These discussions will be counterproductive if the workers are allowed to bring up issues that have nothing to do with the matter at hand or if their concerns are topics from a time that preceded the current administration. Allowing these workers to express themselves is acceptable, but allowing them to dictate the direction of the work group is not.

IMPLEMENTING THE PROCEDURES

This is important to note because in order to establish and maintain the effectiveness of the SOPs, they should be incorporated into the normal operation and maintenance activities. If a lackluster, inconsistently responsive culture is allowed to dictate how the maintenance and operation effort is managed, trying to bring in the necessary improvements these procedures offer will require a major shift in the customs of the existing department/division.

Once the dysfunction has been addressed, or becomes part of the plan to move the productivity of the work group forward, the use of SOPs can enhance the workers' knowledge and understanding of their assigned duties and confirm their value to the impacted workers. When more workers understand their value to the organization, the level of dysfunction can be reversed.

The value of implementing SOPs into the regular operation and maintenance (O&M) processes includes displaying the importance of the procedures in accordance with other vital processes. This connection when continuously presented to staff during development, training, and evaluations will eventually reap dividends in the field crew's effectiveness.

A good method of assisting in the transition toward the use of SOPs is to allow them to be accounted for within the duties of the field crews and work teams responsible for their engagement.

An example of this is in the number of valves a crew is required to operate each day, week, month, etc. When addressing the crew, a review of the SOPs and associated documents should identify a certain amount of time to complete the assignment, resulting in a certain number of these tasks being completed each day, week, month, etc. This approach directly connects the SOP with the proper completion of the assigned task(s).

Another straightforward example is establishing the total lineal feet of sewer main to be cleaned per day, week, or month by crews performing this vital task. Once again, a review of the steps of the SOP and associated documents, addressing all questions and/or concerns, should be done to clarify any issues. These goals should be directly linked to any regulatory requirements associated with the agency maintaining the system.

Frequently, when new initiatives are presented, they are brought in as a separate entity or an additional duty without connecting them to or incorporating them into what is currently being done.

This should be presented in a manner that shows how these procedures for accomplishing the required duties are beneficial because of their efficiency over the previous approaches, their consistency, and their alignment with industry standards.

They should be presented as a better way of getting the work done. It is somewhat similar to the way existing service lines used to be replaced by "open-cutting" the trench. In some cases, each crew had its own way of setting the job up, performing the tasks, etc. However, when the use of "insitu" tools came along, it allowed workers to replace the service line with much more consistency and efficiency, and fewer excavations, as well as a reduced time of disruption to the customer.

The field crew and even the agency could have continued to use the traditional open-cut method but eventually would have realized the challenges and additional expense of maintaining those outdated techniques.

When SOPs are presented as the "approved" processes, a shift in what is expected should be made as well. This connection shows workers the path for the transition taking place related to how the work is

IMPLEMENTING THE PROCEDURES

to be performed. This transition is critical and must be monitored to guarantee that all impacted field crews and work teams responsible for their utilization are on the same page with the expected goals.

As mentioned previously, the implementation approach taken should consider the ability of the impacted workers to absorb the information presented. To maintain consistency, the SOPs should be presented to the group responsible for using them in the same manner in which they were developed.

This requires the agency to be mindful of the optimum approach to align the development and implementation phases.

A process of presenting the established procedures prior to final implementation should be identified in order to gain the full value of the effort and reduce any deterrents to their success. The existing capacity or workload of the review teams is vital. The other duties these workers are responsible for should be taken into consideration as part of the timeline for completion. This is intended to prevent the potential for delays and/or failures in the delivery of the desired effort.

The process should incorporate a schedule for having the procedures reviewed by selected field personnel/crews to determine their accuracy, practicality, consistency with industry standards, and potential for effective execution.

The implementation phase is a great time to evaluate the crew's acceptance of the procedures, to determine an adequate modification schedule, and to confirm the knowledge level of those responsible for using the procedures.

Making sure all impacted workers understand the approaches taken to implement the utilization of the SOPs enhances the potential for success.

Informed workers are better at agreeing to change and tend to be more compliant. In contrast, employees who feel they were "put upon," without the benefit of knowing what is going on, are more prone to rebel either passively or, in some instances, openly.

Allowing time for employees to express their concerns, even "pushback", should be encouraged to demonstrate management's interest in their opinions. All concerns should be addressed to convey their value. Some may not be able to be resolved initially and should be identified for future discussion; however, they should all be addressed, even if it is to express that what was presented cannot be dealt with within the structure of the process. This prevents the possibility of the one concern or "sticking point" later disrupting the intended success.

Conveying the development/implementation plan up front also can potentially reduce the amount of time necessary to deliver the SOPs. Delays are often a result of ambiguity within what was presented/ expected.

When there is clarity in what is presented and contrasting discussion is allowed, the shorter path to agreement usually follows.

Establishing time frames for implementation, training, and assessment provides a level of accountability necessary to complete the process. These time frames should be reasonable and based on what has been accepted by all involved, considering the participating workers' other existing duties.

If the identified schedule is maintained, resulting in the implementation effort being fulfilled, the employees' trust in the process as well as in the agency can be reinforced. This alone will convey the value intended for the process and the SOPs produced.

Additional positive consequences associated with the implementation

IMPLEMENTING THE PROCEDURES

phase are that they offer a great time to evaluate the field crews' acceptance of the procedures to facilitate an adequate modification schedule, and to validate the knowledge level of those engaged in their use.

SOP implementation should provide benefits for maintaining an efficient operation, be user-friendly, and ultimately impact the customers and the environment in a positive way.

Questions to Consider

What should be monitored when implementing SOPs into the O&M process?

Describe the benefits and challenges of introducing SOPs before full implementation.

Identify considerations for introducing the time frame necessary for the implementation of a storage facility inspection SOP.

Identify critical points to be discussed with disgruntled and/or underperforming workers regarding the implementation of an SOP.

CHAPTER 7

Testing and Evaluating the Procedures

THE TESTING AND proving processes benefit the employees involved in their development, the agency's presence in the community, and ultimately, the customer.

Once preliminary procedures have been developed, they should be tested to guarantee the desired expectation prior to final implementation. Testing these processes involves a balance between what is intended in the hoped-for efficiency, productivity, and effectiveness of the SOP and the workers' ability to understand and achieve the anticipated goals.

The testing and evaluation of SOPs offer the best opportunity to provide effective procedures. In order to guarantee a fair and objective assessment of the preliminary SOPs, those performing the tasks should be briefed on the importance of their role. They should also be encouraged to follow the steps identified and to accurately document any confusion or difficulty encountered. This helps produce the desired useful SOPs.

Testing and evaluating the processes for effectiveness should involve the desire to improve productivity. This goal should not be pursued at the expense of other critical objectives but in combination with them. For instance, critical steps to maintain the safe execution of the SOP should not be sacrificed to decrease the time to complete the tasks.

The demonstrated understanding of what is essential to cost-effectively perform the testing/evaluations should be part of the criteria for those completing these assessments.

The process should consist of reviewing them against industry standards. These standards should obviously align with the goals of the organization. They should be comparable to those currently used by the reviewers or at least consistent with what they are familiar with.

If newer processes are presented within the body of the SOP, they must be thoroughly understood by the reviewers in order to guarantee an equitable assessment.

Notes should be captured during the testing and evaluation process to reflect the ease and/or difficulty of performing the steps identified, the times associated with their completion, and the quality demonstrated as part of their conclusion.

Part of these notes should reflect any anomalous activities. Some of these activities experienced by field crews and work teams reflect additional time required to address an unusual situation, such as more time than normal to stage a work zone, a particular difficulty at the jobsite that consumed an inordinate amount of time, a simple repair that morphed into a much bigger job, etc.

The accurate capturing of these issues and the durations associated with their processing are essential to properly achieve the established goal of producing SOPs with reasonable intervals for their completion.

TESTING AND EVALUATING THE PROCEDURES

An evaluation based on the skill level of the employees is essential for a number of reasons. First, as mentioned before, it displays their awareness of the requirements they are expected to know. Secondly, it provides a mechanism to use in scheduling the improvement of underperforming field crews and work teams. Thirdly, the SOPs display the progression of employee knowledge, which the agency is pursuing.

To be clear, an underperforming field crew is not qualified to evaluate potential SOPs and should not be permitted to do so. If they are used, there is a risk of short-circuiting the anticipated success. Criteria for the type of employees and/or work teams to be used in the development, implementation, and testing phases were covered earlier in this document.

A fair and equitable evaluation can furnish information requisite for determining the best avenue for developing workers, along with the encouragement needed to accept operational enhancements.

If possible, the team(s) testing the procedures should be different from those used to develop them. This offers an objective view of the established processes from an alternative perspective of how they will be used regularly. This effort can often aid in possibly capturing missing critical steps within the procedure or identifying necessary but overlooked components.

The knowledge and skill level of those evaluating the procedures should be based on their demonstrated understanding of what is essential to fulfill all aspects of the assignment(s).

Another consideration is to have the procedures evaluated by a newer employee or novice to determine whether the instructions are clear enough to be understood by a worker with minimal experience. This approach offers value when there are a greater number of employees to be developed and/or trained.

DEVELOPING EFFECTIVE STANDARD OPERATING PROCEDURES FOR WATER UTILITIES

The result of an appropriate evaluating/testing process yields vetted, ready-for-final implementation SOPs, or recommendations for enhancement to make the procedures better.

The next step is to identify a schedule for placing the final set of SOPs into use or for determining a plan to apply the recommendations for enhancement and then to repeat the testing/evaluation until the final version is completed.

One of the best ways to determine how effective the performance of the steps of an SOP is, is to first identify a reasonable amount of time it should take to complete them. This reasonable time should consider all of the activities within the SOP, including associated actions such as travel time, prep time, and other related duties, based on the goals of the organization.

These times should be established initially based on discussions with the development team members and then adjusted when more accurate information has been gathered. As mentioned before, they should not be adjusted frequently but based on a predetermined schedule of review and modification.

The discussions to establish the initial times for completing the steps of the SOPs are vitally important and should be done diplomatically. It may be advantageous to accept longer times initially with the intent of reducing them as the reviewing field crews and/or work teams become more familiar with what is expected.

The captured information from the testing and evaluating phase should be reviewed with the goal of confirming the initially accepted time period and/or gathering data to confirm its need to be adjusted.

Certain components present the need for time modifications within the SOP. The removal of asphalt versus concrete as part of an ex-

TESTING AND EVALUATING THE PROCEDURES

cavation is one example. Working on a residential street versus a major roadway is another. The majority of the steps within the SOP are essentially performed the same way in either locale or scenario; however, more time has to be allotted for one over the other.

Another important item that tends to add time to the completion of SOP tasks is the increased depth necessary to come in contact with the asset at one location compared to another. Once again, same steps for the same or similar job; however, an anomalous issue (increased depth) adds time to how long it will take to finish the work.

These anomalies should be identified in the job notes of the SOP to further explain why it took a lot longer to complete the steps. Reviews of these anomalies help to determine the reasonably accurate amount of time to perform assignments.

Some agencies have gone as far as providing factors to their completion times to account for these anomalous occurrences. These factors provide information to balance the reasonable times against which field crews and work teams will be evaluated.

The frequency of the testing and evaluation processes should be consistently determined to make sure all aspects of the procedure are in alignment with the intended goals. The more related information considered, the more reliable the processes and the determined times associated with them should be.

DEVELOPING EFFECTIVE STANDARD
OPERATING PROCEDURES FOR WATER UTILITIES

Questions to Consider

Identify criteria for appropriately testing an SOP to install an inline gate valve.

Identify the steps essential for testing a recently implemented SOP for cleaning a sewer main.

Describe the procedures needed to correct an SOP that failed the testing process.

Identify the impact the skill level of the employees has on testing and evaluating SOPs.

Provide the benefits of accurately capturing notes when testing and evaluating an SOP.

CHAPTER 8

The Components of Standard Operating Procedures

THE COMPONENTS OF SOPs should succinctly convey the title, purpose, process, and other vital attributes of the task in a manner that is easy to understand and follow.

The contents of SOPs may vary based on the approach the agency takes to convey information. The document can potentially consist of the purpose of the procedure; the intended application or scope; necessary acronyms and/or definitions; a potential listing of equipment used in the process; identification of any applicable local, state, or federal regulations; easy-to-follow instructions; and possibly attachments or pictures to further explain the procedure.

In the statement of its purpose or goal, the standard should convey its value to the system as well as to the organization. The purpose of the procedure reflects its importance to the sustainability of the system. When this importance is communicated in this part of the SOP, it directly connects the value of the procedure with the expressed steps to be performed.

The purpose statement can also reflect the desired efficiency to be achieved. With particular SOPs, it is essential to convey the standard as offering a more reliably efficient approach. This may not apply to every SOP, but to the ones that it does, it can express these added benefits to the system.

As mentioned earlier, an orientation to the safe execution of the steps identified within the SOP should be a major factor. It may be difficult to address safety within the purpose statement or goal of the document; however, its importance should at least be implied.

The overall purpose of the standard should reasonably convey its plan to provide a sustainable response to the issue it addresses. The continuity of purpose and process can be expressed in a way that offers a clear understanding to its readers, especially newer workers.

The potential considerations of the purpose or goal declaration involve being clear and concise; connecting the statement to the benefit of the system, organization, or community; and aligning the formal notice portion with the accompanying procedures.

It should be the intent of the document drafters to make sure the SOP's purpose or goal statement is not extremely wordy or rambling. Wordy purpose statements tend to lose their focus or miss their attempt to convey the intended plan. It should not isolate the purpose from the value to those it is intended to serve.

The following examples of purpose and goal statements offer basic content for the identified procedures:

Purpose: The intent of this standard is to provide steps for the investigation of water quality concerns presented by customers with a focus on maintaining the operational integrity of the system.

Purpose: The purpose of this procedure is to provide a course for the

THE COMPONENTS OF STANDARD OPERATING PROCEDURES

repair of water main breaks to maintain the quality of the water in the distribution system.

Goal: The goal of this procedure is to provide a plan for the flushing of fire hydrants to maintain the water quality of the system; aid in the suppression of fires; and capture flow, pressure, and other operations data.

Goal: An effective valve exercising program is essential to improve customer service, maintain distribution system reliability, and ensure water quality control.

The development of the intended application or use should make the process easy to understand. It should offer a solution-oriented approach, coordinating the value of intent with the procedure.

The necessary acronyms and/or definitions should be industry consistent. If there is ambiguity about the acronyms intended to be used, they should either be explained or not used in that form. These acronyms should be process effective or directly applicable to the procedure in which they are written. More importantly, they should be known and understood by the employees using them.

The benefits of acronyms are that they can

- save space in the body of the SOP,
- offer industry-understood meanings, and
- enhance the knowledge of the workers.

The challenges of using acronyms are as follows:

- If not industry-consistent, they can confuse the intent of the document.
- If their meaning changes, a modification of the SOP will be required.

DEVELOPING EFFECTIVE STANDARD OPERATING PROCEDURES FOR WATER UTILITIES

Depending on how the SOPs will be developed, written, and used, there should be consideration for having a list of the equipment required for the process. If equipment and materials are identified within the SOP, they should reflect their consistency with industry standards.

Their reference should comply with the process identified and be familiar to the staff responsible for their use. If staff members are not reasonably familiar with the equipment and/or materials referenced in the SOP, they should not be listed, or steps should be taken to update the workers' knowledge. Their presence within the document should have an orientation toward efficiency and safety.

The equipment and/or materials listed in the body of the SOP should offer the best technology for the solution sought. Outdated or obsolete equipment or materials should not be referenced or used. If there are concerns regarding listing equipment and/or materials that could potentially be subject to change, their use in the document should be evaluated based on that possibility.

The identification of applicable local, state, or federal regulations within the body of the SOP reflect a value and understanding of the referenced requirements. Their reference in the SOP produces an implied adherence to the identified requirements. Their presence also potentially demonstrates the organization's commitment to the regulation(s) listed. Requirements are always changing, and for that reason, when listed, the reference should specify the latest version of the directive, or they should be consistently monitored to reflect the current version.

Attachments and pictures added to the SOP often aid the visual learner in his or her desire to understand the process. Some workers learn by seeing, and pictures showing some of the steps of the procedure can be valuable. Pictures can confirm the particular steps identified. They remove some degree of ambiguity from the procedure and should be updated as the technology advances.

THE COMPONENTS OF STANDARD OPERATING PROCEDURES

The steps listed within the body of the standard should accurately convey what is necessary to fulfill the objectives in an orderly manner. Though the steps should be the most straightforward component of the SOP, consideration should be given to presenting them in a manner that displays the desired efficiency. As mentioned before, they should be clear, easy to follow in a safe manner, and in compliance with all safety requirements of the organization.

Some examples of water distribution system tasks to be standardized are as follows:

Fire hydrant assessments and installation	Distribution valve operation, maintenance, repair, and replacement
Fire hydrant operation, maintenance, rehab, repair, and replacement	Water distribution main assessments
PRV operation, maintenance, repair, and replacement	Water distribution main installation, repair, and replacement
Service line installation, repair, rehab, and replacement	Water main disinfection
Water meter assessments	Transmission main assessments
Water meter repair and/or replacement	Transmission main installation, repair, rehab, and replacement
Water storage facility inspection	Transmission main valve assessments
Booster pump station maintenance and operation	Transmission main valve installation, rehab, repair, and replacement

DEVELOPING EFFECTIVE STANDARD OPERATING PROCEDURES FOR WATER UTILITIES

Some examples of wastewater collection system tasks to be standardized are as follows:

Gravity sewer main assessments and/or inspections	Lift station assessments and/or inspections
Gravity sanitary sewer main installation, maintenance, rehab, repair, and replacement	Sanitary sewer lift station operation and maintenance
Gravity sanitary sewer main cleaning	Sanitary sewer manhole assessments and/or inspections
Sanitary sewer lateral installation, repair, rehab, and replacement	Sanitary sewer manhole installation, maintenance, rehab, repair, and replacement
Sanitary sewer force main assessments	Sanitary sewer accesses and easements maintenance
Sanitary sewer force main installation, repair and/or replacement	Performing sanitary sewer flow test, etc.
Removing blockages from sanitary sewer laterals	Performing sanitary sewer dye test, smoke tests, etc.
Removing blockages from sanitary sewer mains and manholes	Performing odor control and root control procedures for sanitary sewers
Trunk sanitary sewer main assessments	Performing I/I assessments for sanitary sewer mains and manholes

Questions have been raised whether SOPs are more appropriate than checklists in some situations. SOPs offer directional procedures associated with a purpose and/or goal. They often connect the importance of the procedure with the processes necessary to accomplish the stated goal.

THE COMPONENTS OF STANDARD OPERATING PROCEDURES

Checklists are usually reflective of yes/no or pass/fail responses. They either confirm the item was accomplished or not, without any indication of how to perform the identified tasks. Checklists often require additional instructions, particularly for new or inexperienced workers.

It is not beneficial to ask if the new worker checked the box yes or no if they had not been thoroughly trained on what constitutes a positive or negative response for the item under consideration. Checklists have their place and can supplement SOPs within a work group; however, they should never be considered as a replacement for them.

Standard operating procedures are intended to provide value to the agency that develops them properly, uses them appropriately, and updates them regularly to enhance the responses of their workforce for the benefit of the customers they serve.

Another possibility is to combine the value of a checklist with that of an SOP. The following example identifies critical components of the work to be done listed on the checklist.

Checklist items to consider for an inline distribution valve replacement:

1. **As-builts and maps acquired for the identified asset to be replaced:** ☐ yes ☐ no
2. **Valves to be operated to isolate the work area have been identified:** ☐ yes ☐ no
3. **Other utility locates for the impacted work area have been scheduled and received:** ☐ yes ☐ no
4. **Traffic plan submitted and approved for the specific work area:** ☐ yes ☐ no
5. **Impacted customers have been informed of the outages:** ☐ yes ☐ no

The above checklist items should support the completion of the

DEVELOPING EFFECTIVE STANDARD OPERATING PROCEDURES FOR WATER UTILITIES

SOP for replacing an inline valve. The accompanying steps within the SOP should reflect those vital to performing the tasks in an orderly efficient manner. The steps listed below are some of the associated ones essential for replacing the inline valve. The following steps are not a complete list necessary to replace an inline valve but are intended to reflect the valve operation portion of the process identified:

Procedures/Work Steps

The following work steps are recommended for distribution system valve operation:

1. Locate valve to be replaced.
2. Notify impacted customers (as required).
3. Photograph the location, identifying the condition of the site.
4. Check the area for potential hazards and implement needed controls.
5. Establish traffic control as necessary.
6. Pull cover.
7. Clean riser as necessary to inspect valve.
8. Carefully place the valve key on the operating nut of the valve.
9. Exercise valve:
   a. Verify the direction for turning the valve to the **Closed** and **Open** positions.
   b. Assume valve is in the full **Open** position.
   c. Begin **closing valve slowly**, increasing torque as necessary to achieve movement (without exceeding the predetermined *Maximum Torque*).
   d. Count the number of turns necessary to achieve the full **Open** position.
   e. Begin **opening valve slowly**, increasing torque as

THE COMPONENTS OF STANDARD OPERATING PROCEDURES

        necessary to achieve movement (without exceeding the predetermined *Maximum Torque*).
- f. Count the number of turns necessary to achieve the full **Closed** position.
- g. Repeat the Close/Open cycle a minimum of three (3) times, or until the number of turns necessary to open or close the valve does not change.
- h. Record the number of turns, cycles, and Maximum Torque applied.
10. Photograph valve if possible.
11. Record the valve dimensions, condition of the valve, and other pertinent information.
12. Replace cover.
13. Prior to departing, evaluate the location for hazards to people, property, or environment; record findings.
14. Mitigate any hazards discovered and initiate the actions necessary to eliminate those hazards.
15. Photograph the site.

The steps used in an SOP can be simple or more complex based on the capacity of those using them, the specifics of the asset or system, and the goals desired to be accomplished. They should also reflect the importance the steps have to the appropriate completion of the SOP.

As identified earlier, the checklist items usually default to a yes/no response without procedural indication of the associated steps. The procedures necessary to accomplish these checklist items are often scripted in a particular manner, or they can be performed a number of ways without losing efficiency.

The use of the material in this book will aid municipal/utility managers and supervisors in utilizing techniques to capture processes that will standardize work assignments.

DEVELOPING EFFECTIVE STANDARD
OPERATING PROCEDURES FOR WATER UTILITIES

Questions to Consider

Describe the components of a purpose statement for the following that are consistent with the goals of your organization:
a. Sewer lateral replacement   b. Water quality concern response
c. Manhole inspection   d. Bringing a water storage facility on line

Explain when or when not to list equipment in the body of an SOP.

Describe the steps of the SOPs for the following: a. A distribution valve replacement   b. The installation of a sewer manhole

Describe the steps needed for an SOP to inspect a booster pump station.

Describe the steps needed to perform maintenance on a sanitary sewer lift station.

References

ASCE (American Society of Civil Engineers) Report Card for America's Infrastructure, 2017

AWWA (American Water Works Association) Distribution Systems Operation and Management, AWWA G200 Standard

AWWA (American Water Works Association) Installation, Field Testing, and Maintenance of Fire Hydrants, (4th edition), AWWA M17 manual

AWWA (American Water Works Association) Emergency Planning for Water Utility Management, AWWA M19 manual

AWWA (American Water Works Association) Sizing Service Lines and Meters (2nd edition), AWWA M22 manual

AWWA (American Water Works Association) Principles and Practices of Water Supply Operations, (3rd edition)

AWWA (American Water Works Association) Rehabilitation of Water Mains, AWWA M28 manual

AWWA (American Water Works Association) Distribution System Requirements for Fire Protection, (4th edition), AWWA M31 manual

DEVELOPING EFFECTIVE STANDARD
OPERATING PROCEDURES FOR WATER UTILITIES

AWWA (American Water Works Association) Steel Water Storage Tanks, AWWA M42 manual

AWWA (American Water Works Association) Water Distribution Operator Training Handbook, (3rd edition)

AWWA (American Water Works Association) Research Foundation Practices to Prevent Microbiological Contamination of Water Mains and Field Pocket Guide (Project #2610)

AWWA (American Water Works Association) Disinfection of Water Mains, AWWA C651 manual

AWWA (American Water Works Association) Disinfection of Water Storage Facilities, AWWA C652 manual

AWWA (American Water Works Association) Disinfection of Wells, AWWA C654 manual

AWWA, (American Water Works Association) Wastewater Collection System Operation and Management, (1st edition), AWWA G520 manual

Grigg, Neil S, Assessment and Renewal of Water Distribution Systems, Colorado State University, AWWA Water Research Foundation project

USEPA (United States Environmental Protection Agency) Emergency Disinfection of Drinking Water, 2017

USEPA (United States Environmental Protection Agency) Drinking Water Infrastructure Needs Survey, 2015, 2011

USEPA (United States Environmental Protection Agency) Guide for Evaluating Capacity, Management, Operation, and Maintenance

REFERENCES

(CMOM) of Sanitary Sewer Collection Systems, 2005

USEPA (United States Environmental Protection Agency) Planning for an Emergency Drinking Water Supply, 2011

USEPA (United States Environmental Protection Agency) Water on Tap: What You Need to Know, December 2009

USEPA (United States Environmental Protection Agency) Water Factoids, 2009

USEPA (United States Environmental Protection Agency) Innovation and Research for Water Infrastructure for the 21st Century Research Plan, 2007

USEPA (United States Environmental Protection Agency) The Clean Water and Drinking Water Infrastructure Gap Analysis Report, September 2002

USEPA (United States Environmental Protection Agency) Small Systems STEP Guide Series, 2003

USEPA (United States Environmental Protection Agency) Community Water System Survey, 2006, 2000

www.ingramcontent.com/pod-product-compliance
Lightning Source LLC
Chambersburg PA
CBHW050014230526
45470CB00003B/963